ON THE HUNT WITH POLAR BEARS

SANDRA MARKLE

Lerner Publications ◆ Minneapolis

THE ANIMAL WORLD IS FULL OF PREDATORS.

Predators are the hunters that find, catch, and eat other animals—their prey—in order to survive. Every environment has its chain of hunters. The smaller, slower, less able predators become part of the prey for the bigger, faster, more cunning hunters. But everywhere, there are just a few kinds of predators at the top of the food chain. These are the top predators. In and around the ice-covered waters of the Arctic, one of those top predators is the polar bear.

Why are polar bears top predators? For one thing, the adults are big enough to hunt and kill their habitat's big prey, such as walruses. By the time a polar bear is fully grown, it can be as much as 9 feet (2.7 m) long, 5 feet (1.5 m) tall on all fours, and more than 10 feet (3 m) tall standing on its hind legs. A male can weigh as much as 1,200 pounds (544 kg). A female is a little smaller and usually only weighs half as much. But she is still powerful.

An adult's paws are dinner plate-sized—about 12 inches (30 cm) wide. Besides being clawed weapons, such big feet act like snowshoes—perfect for hunting during the Arctic winter. When this hunter swims through open water, its big feet act like paddles.

WOW!
A polar bear can run as fast as 25 miles (40 km) per hour in short bursts.

Of course, a polar bear can't swim forever without resting. During the winter, chunks of ice called ice floes float on the Arctic Ocean. A polar bear can use these as resting sites. Polar bears also use ice floes to hunt for seals, which are their main prey.

Polar bears have three strategies for catching seals on the sea ice. First, this hunter slowly paddles so it seems to be just another ice floe. That way it sneaks up on a seal that's resting on the ice. Or a hunting bear creeps across a large ice floe until it's close enough to charge and catch dinner.

But a polar bear's sneakiest hunting technique is to find a seal's breathing hole. Then the hunter waits as long as necessary for its chance to catch any seal that comes up for air.

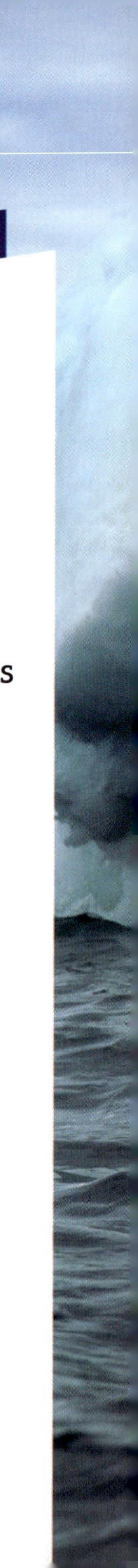

A big reason a polar bear is a top predator is its teeth. And what weapons those are! It has forty-two teeth—ten more than an adult human. This includes four large, sharp canines that easily slice through tough stuff, such as a walrus's thick skin. The rest of the bear's teeth—even its back molars—are sharp-edged, so its bite slices like scissors. The polar bear's bite force is powerful enough to break bones.

A polar bear needs a powerful bite because, unlike other kinds of bears that eat a lot of plant matter, it's a carnivore. It only eats prey animals. And while its teeth are just right for biting, they aren't very good for chewing. So it bites and gulps down chunks. Then its digestive system goes to work to process what it eats. If it gets a chance, a polar bear will eat a lot—up to 20 percent of its body weight in one big meal. No wonder a polar bear naps after it feasts!

A polar bear's body has special features that help make it a top predator in the Arctic. Its fur coat looks as white as its frozen world, so this hunter can blend in and ambush prey. Its coat is also double thick—an oily, water-shedding outer layer over a woolly, body heat-trapping undercoat. So it stays warm even through blizzards or while swimming in icy water.

Polar bear hair has a special feature. While it looks white, all the long, outer hairs are actually see-through, hollow tubes. When the sun is shining, the air inside the hairs warms up, channeling heat to the bear's skin. Below its skin, a polar bear has a thick layer of blubber. This fat provides insulation against cold. Its coat and blubber are so effective at keeping it warm, on a sunny day a polar bear may stretch out on ice to cool off.

WOW!

Besides keeping it warm, a polar bear's blubber makes this massive bear so buoyant that it easily floats.

A polar bear is physically suited to live and hunt in its Arctic Ocean habitat. But it's the bear's senses that really make it a top predator. Its ears high on its head are sound scoops listening for prey. Its eyes see well even in the dim light of the Arctic winter.

WOW!

Each of a polar bear's eyes has a third eyelid, a nearly clear covering that acts like goggles, protecting its eyes from icy bits in the air or water.

A polar bear also has an amazing sense of smell. It can even detect a seal's scent when its prey is inside a den under the snow. When a seal is on top of the ice, a polar bear's super sniffer can smell and track this prey from about a half mile (0.8 km) away.

WOW!
Male polar bears and females that aren't pregnant hunt all winter.

But the number one reason a polar bear is a top predator is because it was trained by another top predator: its mother.

The pregnant female mated sometime between March and June, but the young only briefly developed—then stopped growing. So the female was able to put all her food energy from eating during the summer and fall into building up fat reserves for the winter. She needs that now because, once she's dormant, her babies start growing again. And about two months after she entered her den, the female wakes up just long enough to give birth.

At first, a polar bear cub is no bigger than an average-sized can of soup, and its eyes are sealed shut. It also has no teeth, and its hair is too short to keep it warm. The cub snuggles against its mother for warmth, wiggles until it finds a nipple, and nurses. It continues to nurse and grow for the rest of the winter.

By the time the mother digs open the den to escape in late March or April, her cubs have changed a lot. They're bigger, weighing from 22 to 33 pounds (10 to 15 kg). They have a fuzzy coat, their eyes are open, and they have a set of baby teeth. It's time for the cubs to face their harsh world.

WOW!

Polar bear cubs quickly build up the blubber layer they need to stay warm because a mother polar bear's milk is about 40 percent fat.

Since she missed hunting and eating all winter long, giving birth and producing milk for her cubs to nurse has used up the mother's stored fat reserves. To survive and keep on nursing she needs to hunt—soon!

Once the cubs have had a day or two to practice walking, the mother leads her new family on their first hunt. The cubs' training to become top predators is already starting as they watch her.

The mother sniffs and tracks a seal with her cubs at her heels. When the seal slips into the water, the mother follows, but the cubs stay on the ice. They haven't yet built up their cold-insulating blubber layer. They don't have long to wait, though, before their mother returns with her prey.

Although the cubs mainly get the food they need by nursing, they soon start sampling whatever their mother eats. Eating and sniffing prey is part of learning what scents their keen noses will need to find one day when they're hunting for themselves. Sometimes, like with this piece of whale meat, food also becomes a toy. Today the cubs play tug-of-war. The game soon turns into a wrestling match, which helps the cubs build strong muscles.

Their first summer is the beginning of their training. But it's also a dangerous time for the young polar bear cubs. That's because so much ice melts during the Arctic summer that the mother has to lead her cubs ashore. There the young cubs are at risk of becoming prey themselves.

This year, it's late November before the sea ice forms again and the polar bears head out to hunt on the frozen Arctic Ocean. The mother leads her cubs there, and they begin their most important lessons—how to hunt for prey on the floating ice floes.

As older cubs, they are perfectly adapted for this environment. And their mother teaches them another lesson to stay toasty warm in the extreme cold—roll on snow or ice to clean up after a meal. Clean fur is best for staying warm!

WOW!
One of the seal-hunting tricks polar bear cubs learn is to belly scoot slowly across an ice floe to avoid making warning vibrations that could scare prey away.

Through the winter, the cubs go beyond observing to practicing their mother's strategic moves. They mimic her stealthy sneaking to get close to a seal. And when charging too early warns the seal in time for it to escape, they learn patience is important. By the time the ice is beginning to melt again, the now two-year-old cubs are catching some prey for themselves, even if it's just a wounded seagull.

Even now the cubs are often too noisy and too impatient for their hunts to succeed. But they don't go hungry because their mother shares what she catches. The cubs' hunting skills improve all through that summer and into the winter.

Finally, sometime between March and June, the two-and-a-half-year-old cubs head off to hunt solo. They are now on their own.

The mother is a lone female again. Sometime during those months—March through June—she is ready to mate. When two males fight a fierce battle for her, she mates with the winner.

After mating, the female hunts and eats, building up the body fat she will need to carry her through the winter. When the days grow shorter and colder, snow falls and piles up. Then she digs her den, crawls inside, and becomes dormant. About two months later, the mother wakes up just long enough to give birth. Her new cubs snuggle to stay warm, nurse, and grow bigger.

Come spring, the mother will lead her new family into the harsh world. Now, she'll begin raising a new generation of the Arctic's top predators.

A NOTE FROM SANDRA MARKLE

Even top predators face challenges, and a global change in Earth's climate is challenging polar bears. First, longer warm periods before sea ice returns are annually limiting the polar bears' wintertime hunting season. That means the bears are less fit and less able to go for an extended time without eating. That's most critical for mother bears entering dens with reduced fat reserves to face giving birth and nursing for months before a return to eating.

But even more challenging for a mother polar bear is snow and ice melting sooner and forcing her and her cubs to exit dens earlier. That means the cubs have less time to nurse and grow before facing the world. That is especially challenging for mothers that choose to den on sea ice. When annual ice is thinner, winter storms may cause the ice to break up, turning their den site into a raft dangerously afloat in rough waters.

Photo by: Skip Jeffery Photography

SNAP FACTS

DIET

It mainly eats seals but also eats large prey, such as walruses.

LIFE SPAN

In the wild, it may live as long as thirty years.

RANGE

They live in the Arctic and primarily roam on the ice-covered Arctic Ocean and along its shores.

FUN FACT

Under its white-looking fur, a polar bear's skin is black, and so is its tongue.

ADULT SIZE

Males are larger than females and around 9 feet (2.7 m) long and weigh as much as 1,200 pounds (544 kg).

YOUNG

A female may have as many as four cubs but more often gives birth to only one or two. Cubs are born about two months after the mother enters her den and stay with her, being raised and trained by her, for two to three years.

GLOSSARY

BLUBBER: a thick layer of fat under the polar bear's skin that helps hold in its body heat and helps it float so swimming is easier

CARNIVORE: an animal that only eats other animals

CUB: a young polar bear

DEN: a shelter in a snowdrift a pregnant female polar bear digs to protect herself and her newborn cubs during the Arctic's fierce winter weather

DORMANT: a state in which a pregnant female polar bear's heart rate and breathing slows so much she doesn't need to eat or pee and can stay inside her den sheltered from the Arctic's harsh winter while her cubs develop

FOOD CHAIN: a series of living things where each is dependent on another as a source of food

ICE FLOE: a chunk of floating sea ice

NURSE: to feed on milk from a mother's body

PREDATOR: an animal that hunts other animals

PREGNANT: when a female polar bear has cubs developing inside her body

PREY: an animal that a predator catches to eat

INDEX

Image credits: Exotic vector/Shutterstock.com, p. 1 (design element throughout); Paulette Sinclair/Alamy Stock Photo, pp. 3, 12, 19; by wildestanimal/Moment/Getty Images, p. 5; Paul Souders/Alamy Stock Photo, p. 7; Andyworks/iStock/Getty Images, p. 8; Imagebroker/Alamy Stock Photo, p. 9; Don Johnston_MA/Alamy Stock Photo, p. 11; GM Photo Images/Alamy Stock Photo, p. 13; USO/iStock/Getty Images, p. 15; Noella Ballenger/Alamy Stock Photo, pp. 16, 18; Steven J. Kazlowski/Alamy Stock Photo, p. 20; Michel & Gabrielle Therin-Weise/Alamy Stock Photo, p. 22; Alexey_Seafarer/iStock/Getty Images, p. 23; BIOSPHOTO/Alamy Stock Photo, p. 24; Arterra/Universal Images Group/Getty Images, p. 25; Visual&Written SL/Alamy Stock Photo, p. 26; Nature Picture Library/Alamy Stock Photo, p. 27; © Skip Jeffery Photography, p. 28.

Cover: John Conrad/Getty Images.

THE AUTHOR WOULD LIKE TO THANK DR. TODD ATWOOD, LEADER, USGS POLAR BEAR RESEARCH PROGRAM, ANCHORAGE, ALASKA, FOR SHARING HIS ENTHUSIASM AND EXPERTISE. A SPECIAL THANK-YOU TO SKIP JEFFERY FOR HIS LOVING SUPPORT DURING THE CREATIVE PROCESS.

FOR
LESLIE PRECHTEL AND ALL THE CHILDREN AT WICHITA COLLEGIATE SCHOOL IN WICHITA, KANSAS

Lerner Publications Company
An imprint of Lerner Publishing Group, Inc.
241 First Avenue North
Minneapolis, MN 55401 USA

For reading levels and more information, look up this title at www.lernerbooks.com.

Main body text set in Aptifer Slab LT Pro medium.
Typeface provided by Linotype AG.

Editor: Brianna Kaiser

Library of Congress Cataloging-in-Publication Data

Names: Markle, Sandra, author.
Title: On the hunt with polar bears / Sandra Markle.
Description: Minneapolis : Lerner Publications, [2023] | Series: Ultimate predators | Includes index. | Audience: Ages 8–12 | Audience: Grades 4–6 | Summary: "Polar bears have amazing strength, teeth, and senses that make them one of nature's best hunters. Readers will learn all about these fierce predators, from their hunting behaviors to how they raise their cubs"— Provided by publisher.
Identifiers: LCCN 2021048323 (print) | LCCN 2021048324 (ebook) | ISBN 9781728456287 (library binding) | ISBN 9781728464435 (paperback) | ISBN 9781728462462 (ebook)
Subjects: LCSH: Polar bear—Juvenile literature. | Predatory animals—Juvenile literature.
Classification: LCC QL737.C27 M32165 2023 (print) | LCC QL737.C27 (ebook) | DDC 599.786—dc23/eng/20211013

LC record available at https://lccn.loc.gov/2021048323
LC ebook record available at https://lccn.loc.gov/2021048324

Manufactured in the United States of America
1-50698-50117-1/27/2022